SAN BRUNO PUBLIC LIBRARY
Y0-BZT-611

This Is a Let's-Read-and-Find-Out Science Book®

REVISED EDITION

What the MOON Is Like

Franklyn M. Branley

illustrated by True Kelley

Thomas Y. Crowell
New York

Other Recent Let's-Read-and-Find-Out Science Books® You Will Enjoy

Air Is All Around You · What Makes Day and Night · Turtle Talk · Hurricane Watch · Sunshine Makes the Seasons · My Visit to the Dinosaurs · The BASIC Book · Bits and Bytes · Germs Make Me Sick! · Flash, Crash, Rumble, and Roll · Volcanoes · Dinosaurs Are Different · What Happens to a Hamburger · Meet the Computer · How to Talk to Your Computer · Comets · Rock Collecting · Is There Life in Outer Space? · All Kinds of Feet · Flying Giants of Long Ago · Rain and Hail · Why I Cough, Sneeze, Shiver, Hiccup, & Yawn · You Can't Make a Move Without Your Muscles · The Sky Is Full of Stars · The Planets in Our Solar System · Digging Up Dinosaurs · No Measles, No Mumps for Me · When Birds Change Their Feathers · Birds Are Flying · A Jellyfish Is Not a Fish · Cactus in the Desert · Me and My Family Tree · Redwoods Are the Tallest Trees in the World · Shells Are Skeletons · Caves · Wild and Woolly Mammoths · The March of the Lemmings · Corals · Energy from the Sun · Corn Is Maize · The Eel's Strange Journey

The *Let's-Read-and-Find-Out Science Book* series was originated by Dr. Franklyn M. Branley, Astronomer Emeritus and former Chairman of the American Museum-Hayden Planetarium, and was formerly co-edited by him and Dr. Roma Gans, Professor Emeritus of Childhood Education, Teachers College, Columbia University. For a complete catalog of Let's-Read-and-Find-Out Science Books, write to Thomas Y. Crowell Junior Books, Harper & Row, Publishers, Inc., 10 East 53rd Street, New York, NY 10022.

All photos courtesy of NASA

What the Moon Is Like

Printed in the United States of America. For information address Thomas Y. Crowell Junior Books, 10 East 53rd Street, New York, N.Y. 10022. Published simultaneously in Canada by Fitzhenry & Whiteside Limited, Toronto.

Designed by Trish Parcell
10 9 8 7 6 5 4 3 2
Revised Edition

Library of Congress Cataloging in Publication Data
Branley, Franklyn Mansfield, 1915–
What the moon is like.

(Let's-read-and-find-out science book)
Summary: A description of sights and experiences on the surface of the moon, based on information gathered by astronauts.
1. Moon—Juvenile literature. [1. Moon]
I. Kelley, True, ill. II. Title. III. Series.
QB582.B73 1986 559.9'1 85-47904
ISBN 0-690-04511-5
ISBN 0-690-04512-3 (lib. bdg.)

(A Let's-read-and-find-out book)
"A Harper trophy book."
ISBN 0-06-445052-X (pbk.) 85-45400

What the
MOON
Is Like

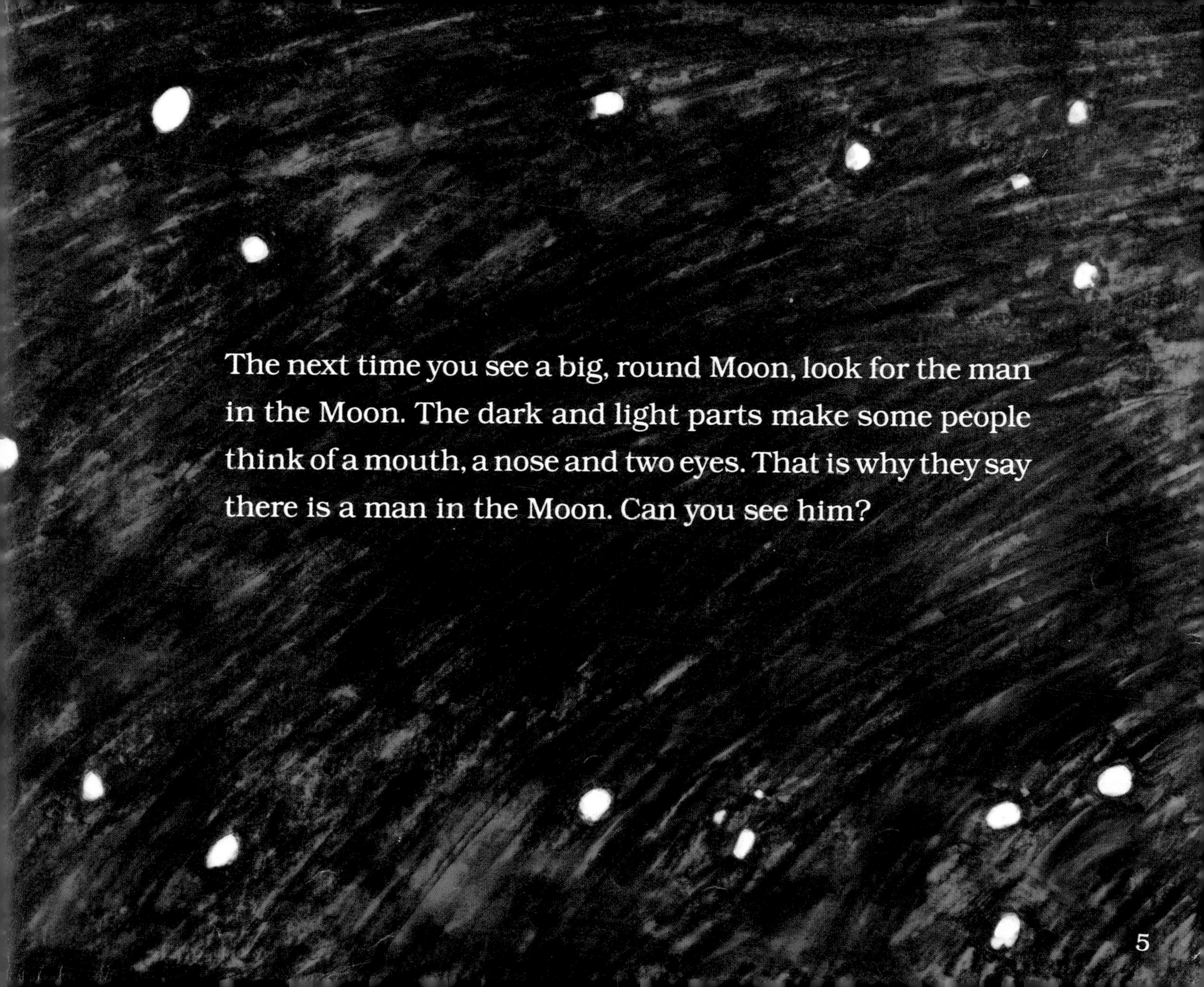

The next time you see a big, round Moon, look for the man in the Moon. The dark and light parts make some people think of a mouth, a nose and two eyes. That is why they say there is a man in the Moon. Can you see him?

Other people say there is a rabbit in the Moon. It has big ears.

Some people see Jack and Jill and their pail of water. Maybe a cat sees a mouse in the Moon.

Next time you see the Moon, look for the man, the rabbit and Jack and Jill. You may be able to see them if you try real hard. Some people try hard but can't see any of them.

If you could see the moon better it would look like this. The light parts are covered with hills, and with holes called craters. Some are many miles across. Others are very small. The holes were made by big rocks that crashed into the Moon long ago.

The dark parts of the Moon are smoother. They are like wide fields. It would take weeks to walk across some of them. They are called the seas of the Moon because they are so flat. There is no water in them. There is no water anywhere on the Moon.

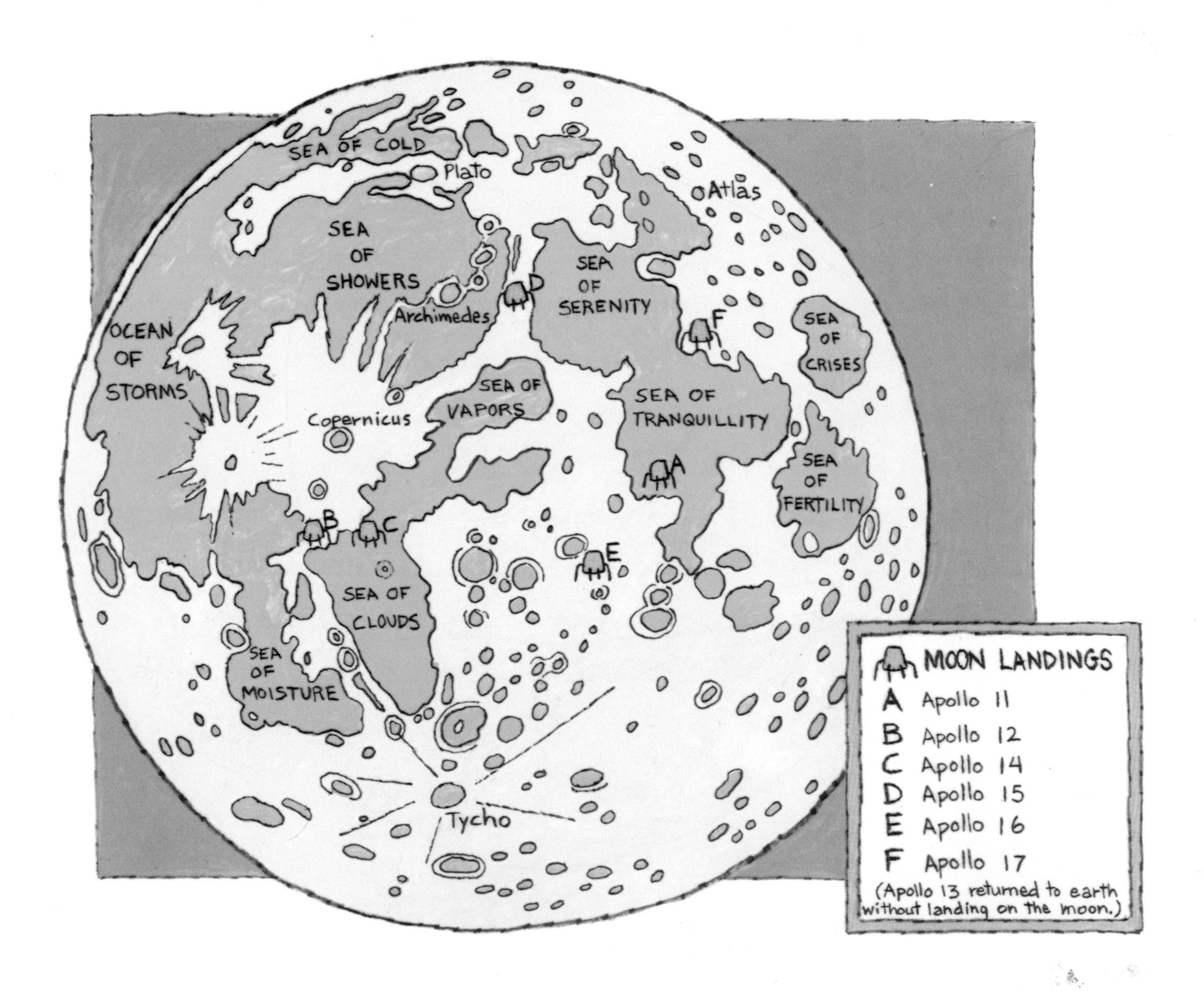
SEA OF COLD
Plato
Atlas
SEA OF SHOWERS
SEA OF SERENITY
Archimedes
OCEAN OF STORMS
SEA OF CRISES
SEA OF VAPORS
SEA OF TRANQUILLITY
Copernicus
SEA OF FERTILITY
SEA OF CLOUDS
SEA OF MOISTURE
Tycho
A
B
C
D
E
F
MOON LANDINGS
A Apollo 11
B Apollo 12
C Apollo 14
D Apollo 15
E Apollo 16
F Apollo 17
(Apollo 13 returned to earth without landing on the moon.)

Twelve men have gone to the Moon. Look at the map to see where they landed. They walked on the Moon. Some of these astronauts rode in a moon car. The astronauts found no air on the Moon. Outside their ship, the men wore space suits. The air they needed was carried inside the suits.

They found small rocks and great big ones. Some were as big as a house. Parts of the Moon are flat. In those places, the astronauts could move quite easily. But there are many mountains and hills. Some are smooth and rounded. Some have large jagged rocks sticking out of them. There are cliffs and deep valleys. The astronauts kept away from them.

The astronauts found no water on the Moon. The Moon is drier than a desert.

They found no living things on the Moon—no animals, no plants, and no fossils of plants or animals that might have lived there long ago.

The Moon is a dead world. It has never had living things on it. It is dead, lifeless and colorless. There is nothing green, blue, red or yellow on the Moon. The whole Moon is drab, dull, grayish brown.

The Moon has a long day and a long night. There are two weeks of daylight on the Moon. Then there are two weeks of darkness.

In daytime the Moon gets very hot. The temperature reaches 250 degrees Fahrenheit. That's hotter than boiling

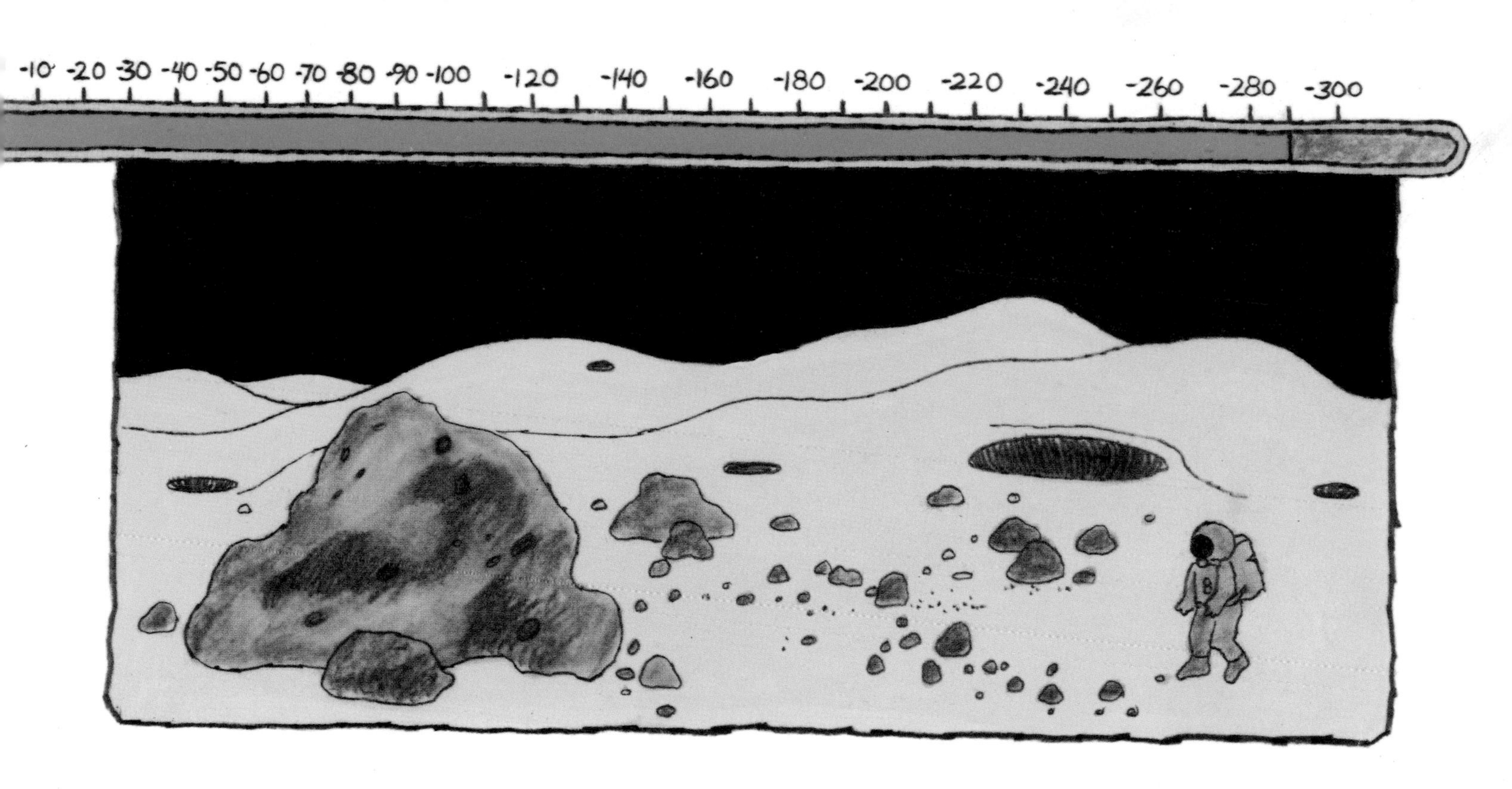

water. It is very cold in shadowed places behind big rocks. There is no air to protect the Moon from the Sun.

At night the Moon gets icy cold, about 290 degrees below zero. There is no air to hold the heat.

Space suits kept astronauts from getting too hot or too cold.

Astronauts liked walking on the Moon. They did not weigh very much because gravity on the Moon is less than it is on Earth. They felt light. You would, too. If you weigh sixty pounds on Earth, you would weigh only ten pounds on the Moon.

Because gravity on the Moon is low, astronauts could take giant steps. They bounded up and down. But their space suits were bulky and awkward. The astronauts had to be careful so they would not fall down.

When they did fall it was hard to get up because of the clumsy space suits, and also because of the moon dust. Much of the Moon is covered with fine dust, almost like powder. In many places the dust is a foot or more deep. When people walk, their shoes make prints in the dust. The moon car had to have wide tires to keep from sinking into it. The dust sticks to shoes and to clothes, too.

The Moon changes very little. It is almost the same today as it has been for billions of years. That's partly because nothing grows on the Moon. Also, there is no water to wear down the hills, and no wind to move dust from one place to another.

Earth is always changing. All kinds of plants and animals live here, and they keep changing. Earth has wind and water. They wear down rocks and move dust and dirt from one place to another.

After the Earth and Moon were formed, big rocks crashed into them. There were also a lot of volcanoes that spouted dust and ash. They spread lava or liquid rock over parts of Earth and the Moon. On Earth much of the lava has been worn away. It is still unchanged on the Moon.

Rocks in space still crash into the Moon. They make moonquakes and new craters. They also crash into Earth, but not very often.

On Earth our daytime sky is bright and blue. That's because water droplets, dust particles and molecules in the air are lighted by the Sun.

There is no air on the Moon—no dust, water droplets or molecules in the sky. There is nothing in the sky to be lighted, so it is always black.

From the Moon, people can see the stars at night, just as we can from Earth. But they can see the stars in daytime, too. In the black sky, they can see stars when the Sun is shining.

When astronauts started for home they could see the Moon just below them. Far away, they could see Earth.

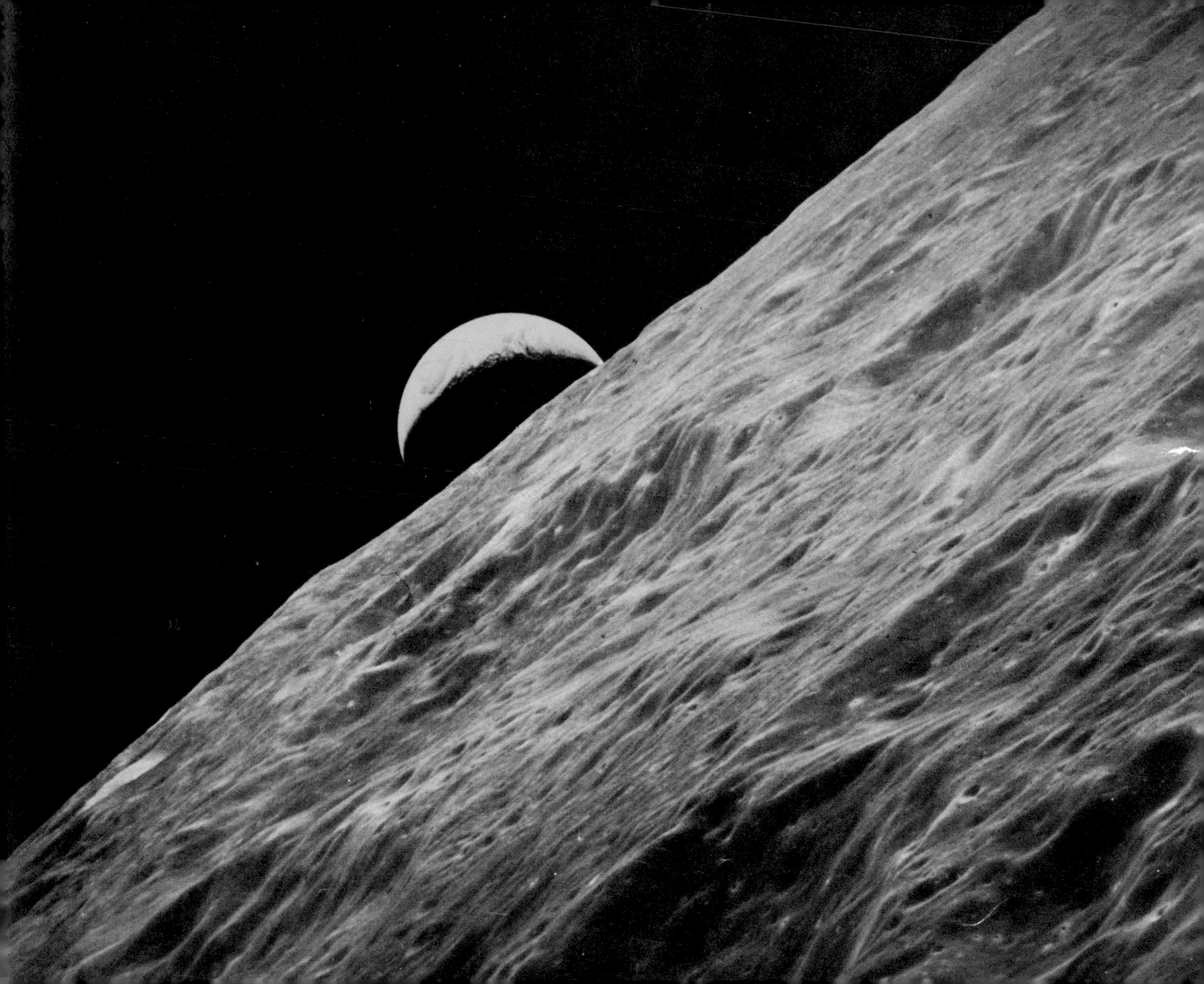

Someday astronauts may go to the Moon again. Once more they will explore it. They will put up buildings on the Moon and live inside them. They may start a Moon colony.

Who knows, someday you may work in a Moon colony. You may be a Moon explorer. Then you'll see for yourself what the Moon is like.

Franklyn M. Branley, Astronomer Emeritus and former Chairman of The American Museum–Hayden Planetarium, is well known as the author of many popular books about astronomy and other sciences for young people of all ages. He is also the originator of the Let's-Read-and-Find-Out Science Book series.

Dr. Branley holds degrees from New York University, Columbia University, and the State University of New York at New Paltz. He and his wife live in Sag Harbor, New York.

True Kelley is the author and illustrator of several children's books. Most recently she has illustrated CUTS, BREAKS, BRUISES, AND BURNS: HOW YOUR BODY HEALS by Joanna Cole and SHIVERS AND GOOSE BUMPS: HOW WE KEEP WARM by Dr. Franklyn M. Branley. She was graduated from the University of New Hampshire and studied art at the Rhode Island School of Design. Ms. Kelley lives in Warner, New Hampshire, with her husband, Steven Lindblom, also an author and illustrator of children's books, and their daughter, Jada.